Saumitra Mukherjee

Extraterrestrial Influence on Climate Change

T0214963

 Springer

Saumitra Mukherjee
School of Environmental Sciences
Jawaharlal Nehru University
New Delhi
India

ISSN 2191-5547   ISSN 2191-5555 (electronic)
ISBN 978-81-322-0729-0   ISBN 978-81-322-0730-6   (eBook)
DOI 10.1007/978-81-322-0730-6
Springer India Heidelberg New York Dordrecht London

Library of Congress Control Number: 2012945109

Printed on acid-free paper

Springer is part of Springer Science+Business Media (www.springer.com)

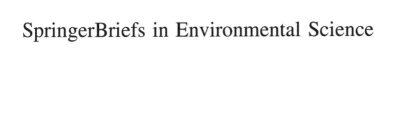

SpringerBriefs in Environmental Science

For further volumes:
http://www.springer.com/series/8868

Preface

Climate changes are being explained by various scientists based on the observational data of last 200 years. Most of the statistical scientific data were not able to answer satisfactorily the erratic behavior of climate. Earth is a tiny part of the Universe and it is linked with the changes of Sun and other celestial bodies. We are continuously facing the changes in the climate due to changes in the Sun and in the intensity of cosmic rays and other collateral changes imposed on the Earth. It has been observed that these changes are taking place before any episodic changes on the Earth. An attempt is made here to simulate the climate change based on the abrupt changes in the atmosphere and geosphere of the earth during solar eclipse. In view of variable Star and Sun the concept of monitoring cosmic ray intensity on Earth has been developed. During solar maximum cosmic rays from the Sun are received in abundance on the Earth. However, it has been observed in various occasion that Sun shows low activities during extragalactic cosmic shower (Forbush effect). An attempt has been made to correlate the changes in the environment of the Earth with the changes in the Sun and extragalactic cosmic rays. Continuous monitoring of the cosmic ray data in different latitude and altitude has already been started across the world with the collaboration of NASA (USA), AOARD (Japan), CRD (Armenia), and JNU (India). This work was initiated in International Heliophysical Year 2007 to correlate the climate change with cosmic ray variables and satellites operating in between Sun and Earth. This book may encourage other researchers across the world who can together be able to identify the missing link to solve the unsolved issues of climate change.

Saumitra Mukherjee

Acknowledgments

The concept of extraterrestrial influence on climate change was developed in my mind more than 35 years ago when I was in my teenage while discussing with my father late Arun Prakash Mukherjee about the change in the Universe and the Earth. Ever changing Earth should be a part of the changes in the Universe; this concept is the key of the sustainable development. If the changes are fast and episodic then the environment is also affected, which may affect the living and non-living components of the environment. I am thankful to my colleagues and students of School of Environmental Sciences, Jawaharlal Nehru University for lending me their moral support during completion of this work. I specially acknowledge Dr. Laszlo Kortvallessey of Hungarian Academy of Sciences and Dr. Milan Radovanovic of Serbian Academy of Sciences. I have worked in association of these scientists in last decade and published with them on various aspects of the Sun–Earth–Cosmic connection. I acknowledge NASA, USA for the use of SOHO satellite data, OMI-KNMI data for the status of aerosol, cloud cover, NO_2, and SO_2 concentration and Yang-Ba-Jing Cosmic ray data available in website. Financial assistance received from AOARD (JAPAN) is also acknowledged. I am thankful to Dr. Dimitar Oozonov, Dr. Barbara J. Thompson, and Dr. Nat Gopalaswamy of NASA USA for their support. I am indebted to Professor Hans J. Hubold an expert of Peaceful Research of Outer Space of United Nations and Professor Ashot Chillingarian of Cosmic Ray Division Armenia for their support. I acknowledge the moral support of my wife Dr. Anita Mukherjee and my children Abhijit and Anisha who cooperated with me and deserve my appreciation.

March 2012   Saumitra Mukherjee

Contents

About the Author

Professor (Dr.) Saumitra Mukherjee (born January 20, 1959 at Varanasi and Ph.D. from Banaras Hindu University in 1989) is an Academician and Scientist with a distinguished career of more than 27 years in Remote Sensing Applications in Geosciences and Space Sciences. He has established interdisciplinary Remote Sensing Applications teaching and research in School of Environmental Sciences, Jawaharlal Nehru University in 1992 after serving in Central Ground Water Board, Government of India (1985–1989), and Remote Sensing Applications Center, Uttar Pradesh (1989–1992) as Scientist. He has worked and published extensively on remote sensing applications and Space sciences in Geology, Hydrogeology, Hydrogeochemistry, Earthquake and other Natural Hazard predictions including snowfall, rainfall and global warming, Land use change, Land subsidence, and Sun–Earth–Cosmic connection. He has published 6 books and 86 papers of international standard. He was awarded as Commonwealth Fellow in 2004–2005 to extend his research in United Kingdom, where he was appointed as Visiting Professor in Department of Earth and Ocean Sciences, University of Liverpool, UK. He has successfully guided 16 Ph.D. scholars and 9 M.Phil. Scholars who are now serving as successful Scientists in Indian Space Research Organization, Space Applications Center, Geological Survey of India, and several Central Universities as faculty members. He is collaborating with NASA, AOARD, ISRO, SAC, GSI, CGWB, DST, MOEF, NIC, and other reputed international and national research organizations.

He has promoted the scientific concept of climate change by extraterrestrial phenomena. His correlation of heliophysical and cosmic variability with global climate change is acclaimed as Mukherjee correlation of climate change.

Chapter 1
Introduction

Climate change has been addressed since last decade based on the influence of human activities like production of industrial effluents, land use changes, and other activities due to development of the society. Studies carried out by geologists of Exeter and Oxford University in 2012 suggests that the carbon in atmosphere was reduced in geological past 470 million years ago by the advent of plant species on the Earth. All these variables might have influenced the environment of the Earth leading to the climate change but the process is slow in comparison to the extraterrestrial influence.

These are very important issues no doubt but the activities due to the influence of extraterrestrial phenomena have not been given its due importance. An attempt is being made here to understand the influence of extraterrestrial activities as one of the important factors of climate change has been attempted here. The influence of Sun and distant stars on the environment of the Earth has been studied during the cyclic changes in the Sun as well as episodic changes in the environment due to the effect of other celestial objects in between Sun-Earth environment. The findings show that if a short-term changes in the Sun-Earth weather due to eclipse can influence the environment of the Earth temporarily it can be possible key of climate change due to abnormal solar behavior.

The study has been carried out based on the changes within the Sun as well as changes during the solar eclipse. During these extraterrestrial changes it has been observed that the Earth changes in its atmosphere as well as geosphere, which may have local effect but the increase of these local effects in large scale may contribute to the climate change. Solar radiation drives atmospheric circulation. Since solar radiation represents almost all the energy available to the Earth, accounting for solar radiation and how it interacts with the atmosphere and the Earth's surface is fundamental to understanding the Earth's energy budget. Solar radiation reaches the Earth's surface by being transmitted directly through the atmosphere or by being scattered or reflected to the surface. One-third of solar (or shortwave) radiation is reflected back into space, while the remaining shortwave radiation at

S. Mukherjee, *Extraterrestrial Influence on Climate Change*,
SpringerBriefs in Environmental Science, DOI: 10.1007/978-81-322-0730-6_1,
© The Author(s) 2013

the top of the atmosphere is absorbed by the Earth's surface and re-radiated as thermal infrared radiation.

The field of global radiation for March month shows a primarily zonal pattern, the radiation decreases with latitude. This occurs because in March, the amount of solar radiation at the top of the atmosphere decreases sharply with increasing latitude. From April through August, latitudinal variations in solar radiation at the top of the atmosphere are less pronounced, because cloud cover plays a strong role in determining the flux reaching the surface. Radiation patterns from April through August are not symmetric. The solar flux is lowest over the Atlantic sector, where cloud cover is greatest. Fluxes peak over central Greenland from May through August. In large part, this illustrates the tendency for the high central portions of the ice sheet to be above the bulk of cloud cover. The highest fluxes are found in June because radiation at the top of the atmosphere peaks in June. Note for June the rather high fluxes over the central Arctic Ocean. This is largely explained in that cloud cover over this region is comparatively limited. From July onwards, radiation fluxes decline. September shows a zonal pattern, which as with March, arises from the strong latitudinal variation in solar flux at the top of the atmosphere for this month.

Recent researches has projected the change in Earths ecosystem due to climate change which needs to be reconsidered based on the parameters of changing extraterrestrial activities of Sun-Earth system. It has been proposed that 2100, global climate change will modify plant communities covering almost half of Earth's land surface and will drive the conversion of nearly 40 % of land-based ecosystems from one major ecological community type—such as forest, grassland or tundra—toward another, according to a new NASA and university computer modeling study. Researchers from NASA's Jet Propulsion Laboratory and the California Institute of Technology in Pasadena, Calif., investigated how Earth's plant life is likely to react over the next three centuries as Earth's climate changes in response to rising levels of human-produced greenhouse gases. Study results are published in the journal Climatic Change. The model projections paint a portrait of increasing ecological change and stress in Earth's biosphere, with many plant and animal species facing increasing competition for survival, as well as significant species turnover, as some species invade areas occupied by other species. Most of Earth's land that is not covered by ice or desert is projected to undergo at least a 30 % change in plant cover—changes that will require humans and animals to adapt and often relocate. In addition to altering plant communities, the study predicts climate change will disrupt the ecological balance between interdependent and often endangered plant and animal species reduce biodiversity and adversely affect Earth's water, energy, carbon, and other element cycles. "For more than 25 years, scientists have warned of the dangers of human-induced climate change," said Jon Bergengren, a scientist who led the study while a postdoctoral scholar at Caltech. "Our study introduces a new view of climate change, exploring the ecological implications of a few degrees of global warming. While warnings of melting glaciers, rising sea levels and other environmental changes are illustrative and important, ultimately, it's the ecological consequences that matter most." When

21st Century Ecological Sensitivity 1

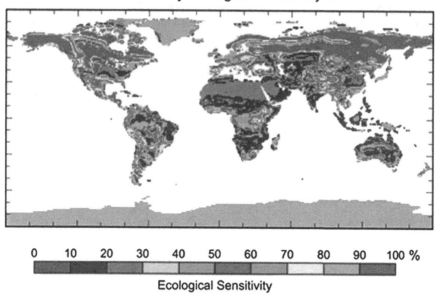

0   10   20   30   40   50   60   70   80   90   100 %

Ecological Sensitivity

Fig. 1.1 Predicted percentage of ecological landscape being driven toward changes in plant species as a result of projected human-induced climate change by 2100. Image credit: NASA/ JPL-Caltech

faced with climate change, plant species often must "migrate" over multiple generations, as they can only survive, compete, and reproduce within the range of climates to which they are evolutionarily and physiologically adapted. While Earth's plants and animals have evolved to migrate in response to seasonal environmental changes and to even larger transitions, such as the end of the last ice age, they often are not equipped to keep up with the rapidity of modern climate changes that are currently taking place. Human activities, such as agriculture and urbanization, are increasingly destroying Earth's natural habitats, and frequently block plants and animals from successfully migrating. To study the sensitivity of Earth's ecological systems to climate change, the scientists used a computer model that predicts the type of plant community that is uniquely adapted to any climate on Earth. This model was used to simulate the future state of Earth's natural vegetation in harmony with climate projections from 10 different global climate simulations. These simulations are based on the intermediate greenhouse gas scenario in the United Nations' Intergovernmental Panel on Climate Change Fourth Assessment Report. That scenario assumes greenhouse gas levels will double by 2100 and then level off. The U.N. report's climate simulations predict a warmer and wetter Earth, with global temperature increases of 3.6–7.2 °F (2–4 °C) by 2100, about the same warming that occurred following the Last Glacial Maximum almost 20,000 years ago, except about 100 times faster. Under the scenario, some regions may become wetter because of enhanced evaporation, while others become drier due to changes in atmospheric circulation. The researchers found a shift of biomes, or major ecological community types, toward Earth's poles especially most dramatically in temperate grasslands and boreal forests and toward higher elevations. Ecologically sensitive "hotspots" areas projected to undergo the greatest degree of species turnover—that were identified by the study include regions in the Himalayas and the Tibetan Plateau, eastern equatorial Africa, Madagascar, the Mediterranean region, southern South America, and North America's Great Lakes and Great Plains areas. The largest areas of ecological sensitivity and biome changes predicted for this century are, not surprisingly, found in areas with the most dramatic climate change: in the Northern Hemisphere high latitudes, particularly along the northern and southern boundaries of boreal forests (as per the latest report of NASA/Jet Propulsion Laboratory Caltech, USA, 2011) (Fig. 1.1).

The ecological sensitivity needs to be remodeled based on the influence of the extraterrestrial activities in Sun-Earth environment.

Chapter 2
Solar Eclipse Influences the Environment of the Earth

The effects of eclipse were observed in a part of China on 23 July 2009 which shows that the cosmic ray intensity decreases with the decrease in Electron flux recorded by the Sun Observatory Heliospheric Observatory (SOHO) satellite. The attempt was based on the changes in the concentration of atmospheric gases like Sulfur Dioxide, Nitrogen Dioxide, Aerosol, and Cloud cover. The paper shows direct correlation of cosmic ray intensity, heliophysical and atmospheric variation during the solar eclipse. It will be useful educational information to understand the atmosphere in variable condition.

United Kingdom BBC weather News UK and NOAA USA.

The Data Interpretation

Air Quality Monitoring and Forecasting in China (AMFIC) addresses atmospheric environmental monitoring over China. The system uses satellite and in situ air quality measurements and modeling to generate consistent air quality information over China. The data downloaded in this paper cover the status of nitrogen dioxide (NO_2), sulfur dioxide (SO_2), aerosol, and cloud cover (http://www.amfic.eu/index.php).

Tropospheric NO_2 Data

Column NO_2 retrievals with the Differential Optical Absorption Spectroscopy (DOAS) technique, and the Koninklijk Nederlands Meteorologisch Institute (KNMI) data combined Modeling/retrieval/assimilation approach. The slant columns from Global Ozone Monitoring Experiment (GOME), Scanning Imaging

S. Mukherjee, *Extraterrestrial Influence on Climate Change*,
SpringerBriefs in Environmental Science, DOI: 10.1007/978-81-322-0730-6_2,
© The Author(s) 2013

Absorption Spectrometer for Atmospheric Cartography (SCIAMACHY), and GOME-2 observations were derived by BIRA-IASB (Belgian Institute for Space Aeronomy), the slant columns from OMI by KNMI/NASA.

SO_2 Concentration in atmosphere:

Sulfur dioxide enters the atmosphere as a result of both natural phenomena and anthropogenic activities, such as coal and oil burning, and SO_2 is therefore one of the atmospheric species related to air pollution. These emissions may have an impact on air quality and therefore it is useful to monitor these compounds. For the Air Quality SO_2 Service, a set of geographic regions covering industrialized China has been defined and each of these regions was monitored. Concentrations of SO_2 were derived from UV measurements by satellite based instruments which are SCIAMACHY, OMI & GOME-2.

Aerosol Data

Global aerosol data are retrieved from GOME and SCIAMACHY measurements. The retrieval is carried out by top of atmosphere (TOA) reflectance's in nadir-viewing conditions, using channels selected in the most transparent atmospheric windows. The algorithms are able to retrieve aerosols over dark-surface as sea-surface.

Level 1B processing level GOME data have been re-calibrated using GOMECal package.

The Fast Retrieval Scheme for Clouds from the Oxygen A-band (FRESCO) cloud algorithm is a fast and robust algorithm providing cloud information from the O_2 A-band for cloud correction of ozone. FRESCO provides a consistent set of cloud products by retrieving simultaneously effective cloud fraction and cloud top pressure. The effective cloud fractions were derived by assuming that the clouds have an albedo of 0.8, and must therefore be interpreted as effective cloud fractions. Note that the derived cloud top pressures are rather insensitive to the assumed cloud albedo. (FRESCO GO-v3 data version) has been used to take into account the ground-pixel cloud coverage fraction. Aerosol classes have been selected among tropospheric aerosol types.

Cloud Data

For the retrieval of tropospheric trace gas columns it is essential to have information on the cloud cover conditions. Both the cloud fraction and the cloud pressure are needed as input for the trace gas retrieval algorithms. The (effective) cloud fraction and cloud pressure are derived from GOME and SCIAMACHY measurements with the FRESCO cloud algorithm. The OMI cloud information is derived with the $OMCLDO_2$ algorithm. The $OMCLDO_2$ Level 2 data product contains the cloud fraction and cloud pressure and ancillary information produced by the OMI Cloud O_2–O_2 algorithm.

Cosmic Ray Data

Downloaded from Yangbajing International Cosmic ray laboratory. The YBJ International Cosmic Ray Observatory is located at in Yangbajing valley of Tibetan highland, a site chosen for its high altitude. This site was selected to compare the atmospheric variables of China and its surrounding area during Solar Eclipse of 23 July 2009. In this observatory, the records used for the present study were mainly cosmic γ-radiation intensity, at an energy threshold of few hundreds GeV, by detecting small size air showers at high altitude with wide-aperture and high duty cycle capability.

Electron Flux Data

Electron flux data were downloaded from SOHO site through www.spaceweather.com site. The electron flux data were measured by the SOHO Satellite in a continuous mode as electron flux. This electron flux plot contains the 5 min averaged integral electron flux (electrons/cm^2-s-sr) with energies greater than or equal to 0.6 MeV and greater than or equal to 2 MeV. Result of correlation of Electron flux data with different atmospheric parameter are enumerated below.

During the total solar eclipse on 23 July 2009, it was possible to correlate some anomaly in atmospheric parameters like cloud cover, SO_2, NO_2, and Aerosol concentration in the air of China and its surrounding areas with the cosmic ray data and electron flux data. Change in electron flux may have changed the ionization of the atmosphere which further gets interrupted by cosmic rays. These mechanisms can lead to the formation of condensation centers. But at the same time interaction between cosmic ray and the atmosphere might change transparency of the atmosphere and the atmospheric temperature also. Calculations of changes of the condensation growth rate of water drops due to changes of the temperature might have been performed. Gradual concentration of Aerosol data has been recorded from 20 to 23 July 2009. Aerosol concentration was low in China–Tibet–Mongolia region, whole China and Japan on 20 and 21 July which found increasing in Guangzhou and Hong Kong areas on 23 July 2009. Gradual rise of cloud cover during the same period has been recorded. Cloud density was observed less intense over Lhasa on 20 July 2009 which was intensified over Beijing and North Korea on 23 July the day of solar eclipse. Slow rise of the Nitrogen Dioxide during the same period has been recorded. Nitrogen dioxide in troposphere column was of very low density on 20 July over Mongolia and north western part of China which was dramatically increased in its highest peak over Beijing on 23 July 2009. Only

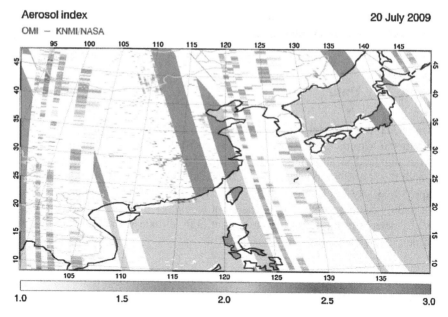

Fig. 2.1 Aerosol concentration over China on 20 July 2009 measured by Ozone monitoring Instrument showing low concentration of aerosol before the solar eclipse (*Source* KNMI NASA)

exception has been observed in the case of SO_2 which shows low concentration from 20 to 23 July. Sulfur dioxide was found high peak on 20–22 July 2009 over Xining and Lanzhou region of China, which reduced on 23 July 2009 (Figs. 2.1, 2.2, 2.3).

Cosmic ray data show a peak value of 5,420 on 20th July but it reduces to 3,850 on 23rd July on the day of Solar Eclipse. Sympathetic to the Cosmic ray data the electron flux data shows a low value on 23rd July 2009 with higher values on 20–22 July 2009 and 24–26 July 2009.

Observational data suggests that episodic change in the Sun during eclipse may influence the terrestrial climate locally. Direct correlation of Cosmic ray intensity and Electron flux in Sun-Earth environment has been done with the Aerosol, Cloud cover, NO_2, and SO_2 concentration in atmosphere. The findings that: During Solar eclipse Cosmic ray intensity decreases locally. During Solar eclipse Electron flux decreases. Cosmic ray intensity and Electron flux has a role in deciding the Aerosol concentration, Cloud cover, NO_2, and SO_2 concentration in terrestrial atmosphere on a local scale. It will be essential to monitor the global cosmic ray intensity in different location of the earth along with the variation of electron flux on regular basis to correlate the aerosol, cloud cover, NO_2, and SO_2 concentration during non eclipse days also. The correlation may be useful to infer the variation of these atmospheric components with the changes in the cosmic ray intensity and electron flux anomaly. This observation can be replicated in various educational institutions for further correlation of atmospheric changes in space and time.

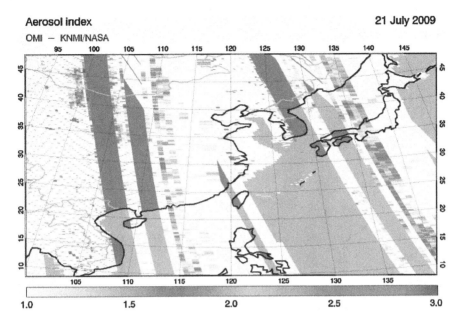

Fig. 2.2 Aerosol concentration over China on 21 July 2009. The Ozone monitoring instrument shows slow increase of aerosol concentration before approaching eclipse (*Source* KNMI NASA)

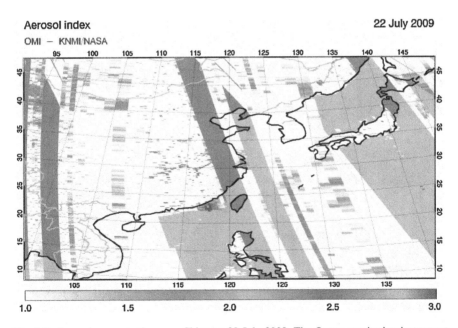

Fig. 2.3 Aerosol concentration over China on 22 July 2009. The Ozone monitoring instrument shows concentrated aerosol formation in China, one day before approaching eclipse (*Source* KNMI NASA)

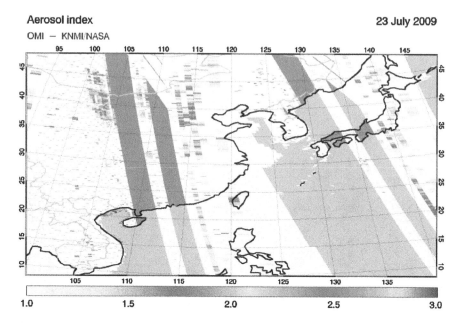

Fig. 2.4 Aerosol concentration over China on the solar eclipse day 23 July 2009 (*Source* KNMI NASA)

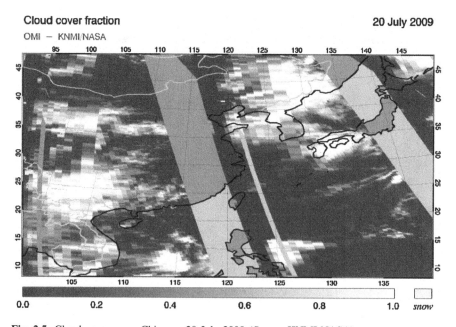

Fig. 2.5 Cloud covers over China on 20 July 2009 (*Source* KNMI NASA)

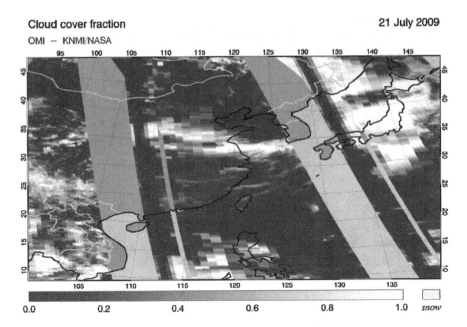

Fig. 2.6 Cloud cover over China on 21 July 2009 (*Source* KNMI NASA)

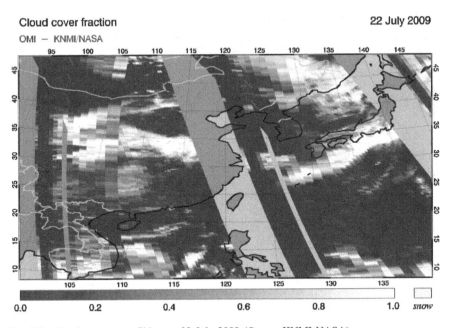

Fig. 2.7 Cloud cover over China on 22 July 2009 (*Source* KNMI NASA)

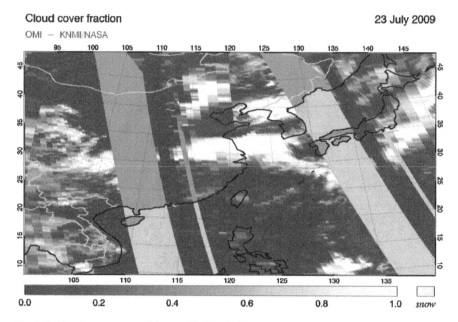

Fig. 2.8 Cloud covers over China on 23 July 2009 (*Source* KNMI NASA)

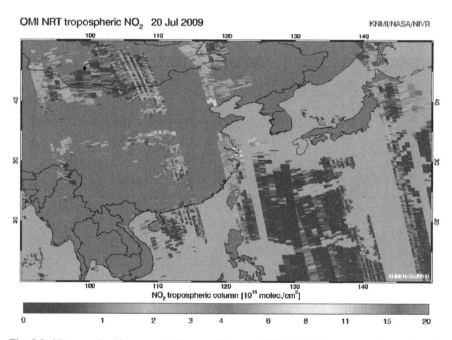

Fig. 2.9 Nitrogen dioxide concentration over China on 20 July 2009. It appears to be very low in comparison with the 23 July 2009 (*Source* KNMI NASA)

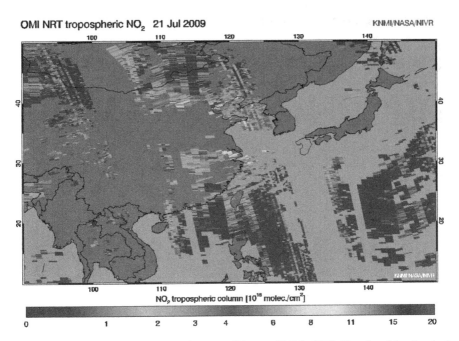

Fig. 2.10 Nitrogen dioxide concentration over China on 21 July 2009. Showing rising trend of NO_2 (*Source* KNMI NASA)

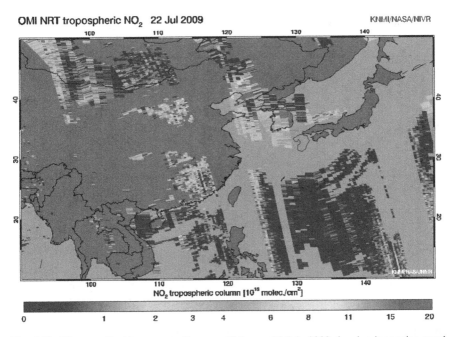

Fig. 2.11 Nitrogen dioxide concentration over China on 22 July 2009 showing increasing trend (*Source* KNMI NASA)

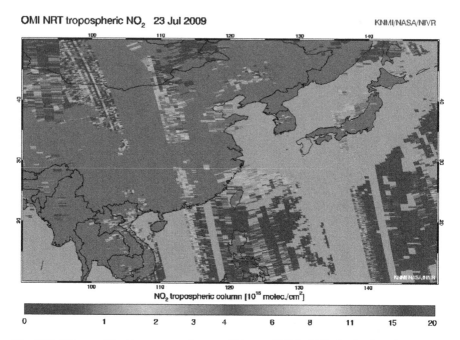

Fig. 2.12 Nitrogen Dioxide concentration over China on 23 July 2009. On the day of eclipse the concentration of NO_2 was higher (*Source* KNMI NASA)

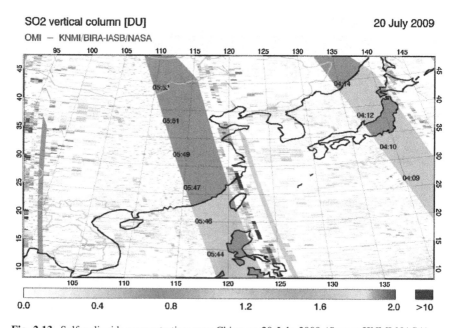

Fig. 2.13 Sulfur dioxide concentration over China on 20 July 2009 (*Source* KNMI NASA)

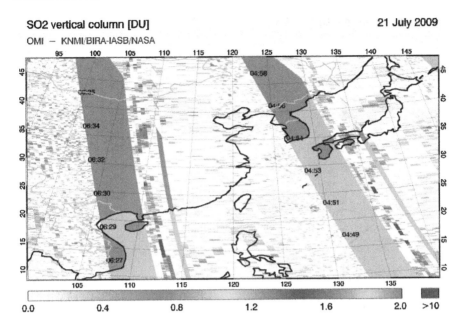

Fig. 2.14 Sulfur dioxide concentration over China on 21 July 2009 (*Source* KNMI NASA)

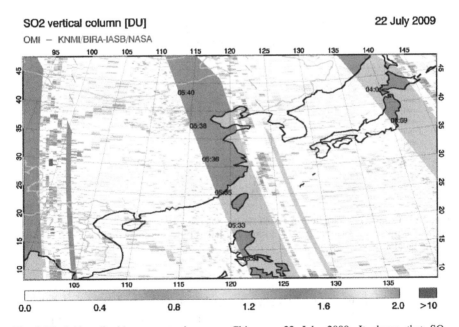

Fig. 2.15 Sulfur dioxide concentration over China on 22 July 2009. It shows that SO$_2$ concentration has decreased before the solar eclipse (*Source* KNMI NASA)

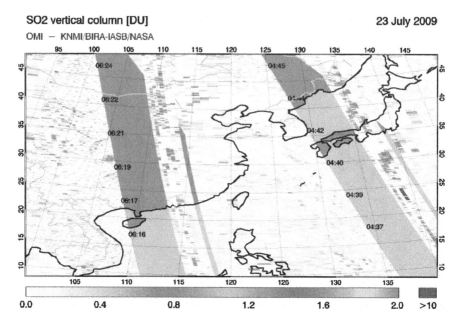

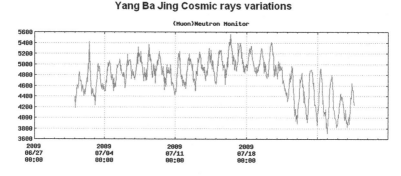

Fig. 2.16 Sulfur dioxide concentration over China shows much decrease on 23 July 2009 (*Source* KNMI NASA)

Fig. 2.17 Cosmic ray variation in China during solar eclipse on 23 July 2009. It shows that during the period of solar eclipse the cosmic ray intensity was decreased (*Source* YBJ Cosmic Ray Laboratory)

Solar spectral irradiance is being measured by Solar Radiance and Climate Experiment (SORCE). It measures the fluctuation of solar variability from 0.1 to 2,400 nm. During the Solar eclipse period the solar irradiance has shown a lower

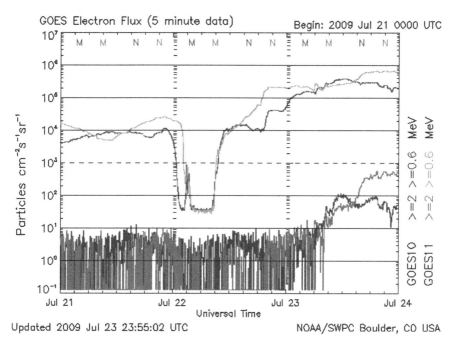

Fig. 2.18 Electron flux lowering during solar eclipse on 23 July 2009. Immediately after the solar eclipse the electron flux shows a rise (*Source* SOHO)

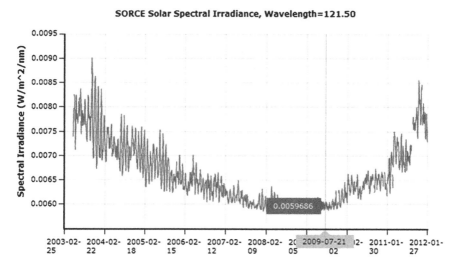

Fig. 2.19 Low solar spectral irradiance recorded during solar eclipse

value which has been found responsible for lowering down the Sulfur Dioxide concentration in the atmosphere (Figs. 2.4, 2.5, 2.6, 2.7, 2.8, 2.9, 2.10, 2.11, 2.12, 2.13, 2.14, 2.15, 2.16, 2.17, 2.18, 2.19).

Chapter 3
Influence of Sun on Atmosphere and Geosphere

It is not possible to study the atmospheric changes in isolation. It has been inferred that there are indirect links existing in between the changes within the Sun with the thermosphere, ionosphere, atmosphere, and geosphere. The Earth's magnetic field acts like a shield, protecting earth from the constant stream of tiny particles ejected by the Sun which is known as the 'solar wind'. The solar wind itself is made of hydrogen atoms, broken into their constituent pieces: protons and electrons. When electrons find routes into our atmosphere, they collide with and excite the atoms in the air. When these excited atoms release their energy, it is given out as light, creating the glowing 'curtains' we see as the aurora borealis (or the aurora australis in the southern hemisphere). Dayside proton aurora spots are caused by protons 'stealing' electrons from the atoms in our atmosphere. The changes in the particle flux depend on changes in magnetic and electric fields. Particle flux from Sun can suddenly change the Kp (planetary indices) which may be pinpointed toward earth in the form of magneto tail. Interaction of the CME particles with the ionosphere, earth's upper atmosphere (between 80 and 200 km above the ground) has been noticed. Scientists doing research with magnetometers just before major earthquakes have serendipitously recorded tiny, slow fluctuations in Earth's magnetic field. Satellites equipped with IR cameras could be used to detect seismic hot spots from space. In fact, when Demeter Ouzounov of the Goddard Space Flight Center (GSFC) examined infrared data collected by NASA's Terra satellite, they discovered a warming of the ground in western India just before the powerful 26 January 2001, quake in Gujarat (SCIENCE@NASA, 11 August 2003).

Correlation of weather fluctuation with Coronal Mass Ejection (CME): To close this considerations let us look at most impressive demonstration of solar activity influence on the lower layers of our atmosphere. The new concepts concerning solar-terrestrial relationship research taken into account the primary processes of the whole Sun-Earth system. Cyclic changes of the general atmosphere circulation are of prime interest as also the transformation and recurrence of circulation forms, which characterize the planetary wave dynamics. The changes

S. Mukherjee, *Extraterrestrial Influence on Climate Change*,
SpringerBriefs in Environmental Science, DOI: 10.1007/978-81-322-0730-6_3,
© The Author(s) 2013

of the atmospheric pressure in the geomagnetically and electronically exited cases (including the solar activity effect) in comparison to the variations in geomagnetically and electronically quite cases (caused by the atmosphere and CME interactions).

The solar-atmospheric interactions were observed more in the Northern India. Delhi, Haryana, Meghalaya Mizoram stretch of India has received more than 100 % from the normal rainfall during the monsoon period of 2003.

The activation of the energetically active regions occurs after the geomagnetic disturbances. The corresponding regions are Delhi–Haryana–U.P–Bihar–Bengal–Orissa–M.P–Meghalaya–Mizoram with high temperature contrasts and a vertical wind shift together with baroclinic instability.

The basic disturbances appear after the geomagnetic ones in the form of the planetary wave structures with wave numbers from 3 to 4. This corresponds to the result about the selective properties of the baroclinic instability and fits the earlier obtained results on the increase of kinetic energy of such waves after the geomagnetic disturbances. Recurrence of draught after 12 years in various parts of the earth due to irregular rain fall is due to change in ionosphere which is triggered by the change in the electron flux during earth directed. Correlation of triggering of earthquakes with CME: In recent years, earthquake prediction has moved from being a guessing game to more of a science. The primary obstacle to understanding earthquakes was geologists' refusal to accept tectonic plate movement (continental drift). Today we know that the crust of the Earth does float on a "sea" of magma, and that earthquakes and volcanism occur primarily in those zones where one plate is rubbing against another—"fault lines." This model is not perfect, because there are often "rogue" quakes, which strike in unexpected areas, but it works extraordinarily well. Nonetheless, there are certain aspects of earthquakes that lend themselves to predicting their incidence—but geologists refuse to pay attention. One factor is that oscillations in the planet's magnetic field often occur right before quakes; these oscillations, as noted above, often correlate strongly with astronomical and solar events. But geologists are afraid to say something ridiculous along the lines of "sunspots cause earthquakes," despite the fact that, in a certain sense, they do… radio wave propagation (e.g. ionosphere "whistlers"), which appears to play a role in earthquakes, and it can be studied by closely monitoring the impact of the solar wind on the ionosphere. We are interested in the sun because of the many influences it has on our lives and our environment. Beyond the obvious considerations of heat and light, some examples of these direct and indirect solar influences are the effects on short-wave radio communications, navigation, use of satellites for communication and navigation, hazards to humans and instruments in space, electrical power transmission, geomagnetic prospecting, gas pipeline monitoring, and possibly weather Seismotectonics and human and animal behavior. During the solar maximum before 11 year it was a sensation that why the change in atmosphere and geosphere is so drastic. Again the Solar maximum has come after 2000. It has been noticed that the number of Sunspots are continuously on rise. Earth directed coronal mass ejections (CME) is very frequent during 2000–2002. Continuous impact of the tremendous amount of energy has frequently changed the Kp (magnetic indices), and free electron in the upper part of the atmosphere. These changes in Sun-Earth environment induced by the Sun have changed the geosphere and Atmosphere from time to time. Occurrence of Earthquakes followed by CME and raised Kp values have sufficient bearing on this hypothesis. All these facts compelled us to rethink and reopen the possible researches in the Sun-Earth environment. Atmospheric perturbations may occasionally preceded earthquakes in different parts of the earth. Different types of earthquake lights have been reported before, during and after severe earthquakes. Some observers have seen red, blue, or white glows, while others have described them as balls of fire or flashes from the sky. Such observations have been assigned different causes. Some attribute them to the lightning from a thundercloud; some to sparks in the electric power lines, while others to the generation of static electricity in the vicinity of focal zone of earthquakes where relative movements of rocks may produce heat and light. Over the sea, such light could arise from luminous marine organisms excited by the vibrations produced by the earthquakes.

Earthquake lights have been reported before Matsushiro swarms during the year 1965–1967 and before some recent earthquakes in China. During the Pattan earthquake (Pakistan) of December 1974, reliable forest officers and doctors far away from the earthquake epicenter observed earthquake lights came from sky. Before the occurrence of Jabalpur earthquake (India) of 1997 a light source came from sky before the occurrence of the earthquake. Experiments have been conducted at the University of Western Ontario, London (Canada) to understand the possible mechanism of earthquake lights. It has been suggested that adsorption of condensation of water could be thought of an energy source for the release of light from solid particles suspended in a cooling column of air above ground. But this theory could not explain the occurrence of light from sky.

Occurrence of lights during earthquake may be explained by the sunspot activities during solar maximum. The Earth has a magnetic field with north and south poles. The magnetic field of the earth is surrounded in a region called the magnetosphere. The magnetosphere prevents most of the particles from the sun, carried in solar wind from hitting the earth. Some particles from the solar wind can enter the magnetosphere. The particles that enter from the magneto tail travel toward the Earth and create the aurora oval light shows. Dangerous particles are not able to penetrate to the Earth's surface but are forced by the magnetic field to move around the Earth. Particles gain entry through the cusps that are shaped like funnels over the Polar Regions or they gain entry far downstream from the Earth. The particles that enter downstream travel toward the Earth and are accelerated into the high-latitude ionosphere and produce the aurora oval light shows. Since the most intense auroras occur at solar maximum, it was once thought that the Sun hurled material out during these raised times of solar activity and that material funneled directly into the polar cusps. However, we now understand that the electrons that cause the auroras come in downstream or from the Earth's magnetic tail. These electrons that enter at the magneto tail are energized locally within the magnetosphere. The solar wind, emanating from the Sun, injects plasma into the magnetosphere and transfers energy to it. Several times a day, the magnetosphere undergoes a disturbance called a substorm. As the substorm grows, most of the solar wind energy is dissipated within the magnetosphere, ionosphere, and upper atmosphere.

This disturbance ultimately causes aurora displays, the acceleration of charged particles to high energies, the emission of intense plasma waves and electromagnetic waves, and the generation of strong ionospheric currents that produce significant changes in the upper atmosphere. These waves and currents often result in severe problems on Earth with communications, power supplies, and spacecraft electronics.

Other higher energy particle radiation that could pose a danger to life here on Earth is forced to drift around the Earth within two large donut-shaped regions called the radiation belts. Invisible magnetic fields are the reason that particle radiation moves in this way. Before the occurrence of Catastrophic Earthquake on 26 January 2001 at Kutch, IMAGE spacecraft captured Gujarat, India an invisible magnetic tail. This is a major precursor of earthquake. More advanced researches should be conducted to identify the geolatitude–longitude which is likely to be affected by the magnetic tail, which may trigger earthquake in active fault areas. Sudden rise in Electron Flux before 36 h from 26 January 2001 (on 24 January 2001) was observed and recorded. An infrared image of the region surrounding Gujarat, India, on 21 January 2001 shows trace of thermal anomalies as Yellow–Orange areas that appeared days before the 26 January earthquake. In various parts of the earth sudden rise in Kp values has been observed before 36–24 h from the occurrence of earthquake. Before the occurrence of 26 January 2001 Earthquake of Kutch, Gujarat, and India Kp values increased from 1 to 5 on 24 January 2001 (36 h before occurrence of 26 January earthquake). Similar observations were observed in several other cases also including Andaman Islands also.

Coronal mass ejection, increase in Kp values (more than 4) (Fig. 2.2), sudden increase in X-ray flux (Fig. 2.3) and electron flux can be forewarning of seismic disturbance in earthquake prone active fault areas and other environmental changes of earth. On 24 January 2001 an earth directed coronal mass was ejected, which took two days to reach the earth surface and a major earthquake of magnitude 7.9 occurred in Gujarat, west coast of India. This area was reported as seismically active in entire world, a total of 65 earthquakes have been reported on the same day (NEIC 2001). Earth directed coronal mass ejection produced a suspected invisible tail of electrified gas. IMAGE spacecraft (NASA Science News 2001) spotted the tail, which streams from earth toward Sun. The region laced by earth's magnetic field, called the magnetosphere, dominates the behavior of electrically charged particles in space near earth and shields our planet from the solar wind. Explosive events on the Sun can charge the magnetosphere with energy, generating magnetic storms that occasionally may affect the active faults in igneous/sedimentary/metamorphic geosphere and change its viscosity (Brandish and Marsh 1989) to trigger the shallow focus earthquake. It may be interesting to observe the series of earth directed CME and occurrence of earthquakes globally. It is not a specular correlation that earthquake of 6.0 magnitude occurred on 11

August 2003 at Andaman Island of Indian subcontinent was preceded by a rise in Kp indices, Electron flux as well as X-ray flux.

It can be concluded that sudden one drop in Kp, Electron flux, Proton flux, and X-ray flux is an indication of atmospheric disturbance. This incidence may be followed by anomalous behavior of Indian monsoon, which ultimately leads to erratic rainfall pattern. Due to low electron flux the local drop in temperature at upper part of atmosphere leads to condense the clouds on the affected part of the earth. Contrary to this the rise in Kp indices, Electron flux, Proton flux as well as X-ray flux leads to occurrence of earthquakes. It is possible that the high particle flux on the earth is responsible for the release of more near infrared radiation before the earthquake as rocks rubbing together—is not responsible for the radiation field (SCIENCE@NASA, 11 August 2003). Position of the Sunspot in Sun is important, coronal mass ejection from the sun may lead to catastrophic changes in a particular part of the planet earth.

A severe geomagnetic storm originated from Sunspot No. 486 on 29 October 2003 has shown its effect on the environment of the earth It began at approximately 1700 UT when a coronal mass ejection (CME) struck our planet's magnetic field. The CME originated by an x10-class explosion from giant sunspot 486. Proton flux, electron flux, and planetary K indices shows sudden rise from 17.00 UTC on 29 October. Proton flux was more than 104 meV, electron flux was jumped to 106 meV while K-indices was at its peak value of 9. The location of Sunspot 486 is found out to be in the southern part of the Sun. The effect of this CME was observed on crustal disturbance in seismically active parts of the earth. An earthquake of 6.8 Richter scale was recorded in the off coast of Honshu, Japan. Before the occurrence of this severe earthquake bright aurora was seen in the sky of Japan. It can be possible to correlate the possibility of triggering of the active fault of Honshu area of Japan by the sudden change in the magnetic shield of the earth due to impact of CME from Sun. The effect of earth directed CME not only triggers the earthquake but it effects the whole environment of the earth which includes destruction of ozone layers to the climate change. Active Sunspots 487, 488, and 492 along with 486 in the southern part of the Sun continues to be a possible threat to the changes in the thermosphere, ionosphere, atmosphere, and geosphere of the earth. In the past first solar flare was recorded on 1st September 1859, largest flare occurred on 2 April 2001. It has been recorded that sunspots and solar flares are regular phenomenon during the solar maximum. Besides the episodic event of earthquake it changes other environmental components including temperature rainfall and magnetic field of the earth temporarily. The correlation of this phenomenon with the environment of earth is being established first time.

Chapter 4
Influence of Sunspots on the Saturn Environment

Influence of Sun can be observed in other planets of the Solar system also.

The Saturn ring has provoked scientists and armature astronomers since the advancement of human civilization. Various hypotheses have been postulated on the connection of Sunspots Earth's environment and changes in the Saturn ring. A concept was given on Planetary-Spin heat, Jupiter-Saturn-Solar, Tidal-Sunspot coupling, and Planetary-Solar alternating magnetic field rings by Chris Landau in 13 September, 2010. It shows that how our planets and stars interact through tidal and magnetic forces.

Three new concepts are proposed

(a) Heating our Earth and the planets by Rotation or Spin, not by radioactive decay.
(b) Jupiter and Saturn's Spring Tides cause the Sunspot cycle and Solar magnetic field reversals.
(c) Planetary rings of gas giants are caused by alternating magnetic fields due to a rapid spin.

The relationships are:

1. Rapid Planetary Rotation causes internal heating, through density increase, changing magnetic and electric flows within the core, resulting in convective heat, continental drift and volcanism.
2. Sunspot Jupiter-Saturn Tidal link for Sunspot cycle timing and Solar magnetic field reversals.
3. Rapid spin of planets and stars (pulsars) leading to alternating magnetic fields, responsible for ionized planetary rings and torus rings. These are suspended halfway between the poles of planets and pulsars. Alternating, pulsing beams of high energy particles from pulsars are due to the alternating magnetic fields within their cores.

S. Mukherjee, *Extraterrestrial Influence on Climate Change*, SpringerBriefs in Environmental Science, DOI: 10.1007/978-81-322-0730-6_4,

Tidal Forces of Jupiter and Saturn may control our Sunspot cycle our two gas giants; Jupiter and Saturn have a large tidal influence on our Sun; much like our Moon has a very large tidal influence on our oceans, our atmosphere, and our solid Earth. These tides where we have competing gravitational forces between the Sun and Moon, cause our daily, measurable influence on our 12 h cycle of the tides on our oceans and large inland seas, our weather, due to compression and expansive forces, and even small earthquakes and minor volcanism changes as documented on Mt Etna in Sicily. The tides on our Sun's atmosphere or photosphere by Jupiter and Saturn affect Earth's atmosphere too. As Jupiter and Saturn revolve around the Sun in 11.85 years for Jupiter and 29.5 years for Saturn, they are either on the same side of the Sun reinforcing the tidal effect (equivalent to the Spring tide on Earth-Moon-Sun system) or on opposite sides of the Sun, also known as Spring tide. This occurs every 10.7 years approximately. Because of Jupiter's mass, the Jupiter-Sun center of gravity position falls 7 % outside our Sun and has a great tidal effect on the Sun. So here we have a simple answer for what causes the pull on our Sun and distorts that atmosphere to create Solar dark spots that are actually magnetic field storms. These storms occur because the Sun is composed of plasma (free electrons, protons, and neutrons moving in circular currents, much like our Earth's atmospheric storms move in circular fashion, driven by heat and the Carioles force. These magnetic storms ionize our Earth's atmosphere and probably provide more condensation nuclei in the form of charged, dust, salt, and smoke particles that either increase or decrease our rainfall over the same 11-year cycle. Planets rotational spin and heat and cool our planet's core to provide the answers to some of the magnetic reversals we see. These magnetic reversals should also be read the other way around and tell us what is happening in our Sun and what is still going to happen. They provide clues to changing events. I think that once we know where all the orbiting bodies in our Solar system are; we will come to understand the big picture that controlled all of these past recorded events; that are written in the pages of our planets. We have only just begun to read the pages in the other planets of our Solar system. Saturn's rings are a magnetic plane of ionized particles suspended halfway between the two poles as the magnetic field oscillates and alternates. This plane is not a torus but a ring (Fig. 4.1).

Saturn observation on 30th June was remarkably noted in view of Sun-Saturn-Earth environment. There was a sharp rise in global temperature from 28 June 2004, which was found on an increasing trend till 2 July 2004. A sharp rise in e-flux, p-flux, and Kp indices were noticed 36 h before CME from the Sunspot no. It has been observed that Saturn ring has been showing some unusual waves in the Cassini latest images. Saturn rings are made up of a mélange of dust-ice-particles, which are influenced by the sudden changes in Sun-Saturn environment. Similar observations are being made on Earth. Thermosphere-ionosphere-atmosphere and geosphere of Earth were found to be changing during coronal mass ejection from the Sun. During the global temperature rise from 28th June to 2nd July the biomass of various parts of Earth were found disturbed. Large numbers of snakes, rats, and other animals living inside the Earth came restlessly on the surface of the Earth.

Fig. 4.1 Saturn ring (*Courtesy* NASA/JPL/Space Science Institute)

Fig. 4.2 Saturn ring full of waves

Table 4.1 Daily geomagnetic data of 28 June–2 July 2004

Date	Middle latitude Fredericksburg		High latitude College		Estimated Planetary		
	A	K-indices	A	K-indices	A	K-indices	
28 June 2004	11	2 4 1 2 2 1 3 3	−1	3 1 1 0 1 1 0 0	13	3 3 1 2 2 2 4 4	High
29 June 2004	15	3 3 3 2 2 4 3 3	29	5 5 5 4 4 5 3 3	20	4 4 4 3 3 3 4 3	High
30 June 2004	8	3 2 3 2 2 1 2 2	21	3 2 3 5 4 4 2 2	10	3 2 3 3 3 2 3 2	Moderate
01 July 2004	9	2 2 3 2 2 1 3 2	24	2 3 5 5 6 3 3 2	13	3 3 3 3 3 3 3 2	High
02 July 2004	−1	−1 −1 −1 −1 −1 −1 −1 −1	−1	−1 −1 −1 −1 −1 −1 −1 −1	−1	2 3 2 −1 −1 −1 −1 −1	Low

On 2nd July onwards the global temperature will decrease as well as it is expected that waves in the Saturn will reduce (Fig. 4.2).

This image shows three density waves in Saturn's A ring. It was taken by the narrow angle camera on the Cassini spacecraft on 1 July 2004 after successful entry into Saturn's orbit. The view shows the dark, or unlit, side of the rings (Table 4.1).

Chapter 5
Cosmic Ray Variation and Environmental Change

Influence of star flare on Sun-Earth environment leading Snowfall Influence of star flares during low Kp and low Efflux condition of Sun-Earth Environment may have produced further sudden drop in magnetic field and Efflux in Sun-Earth environment. This may have caused by repulsion of magnetic field in Sun-Earth Environment and star flares. Time varying ionospheric currents caused by geomagnetic storms originating from the star-Sun-Earth environment induced electron flux in the Earth's atmosphere. Ionospheric current variation has direct influence on atmospheric temperature. On 25 December 2004 hailstorm and snowstorm has been reported in northern hemisphere, while in tropics the temperature dropped suddenly led to foggy and smoggy condition. Star-Sun-Earth atmosphere temperature variation is guided by the respective position of star and its influence on the geomagnetic co-ordinate of the Earth.

Our planet is a part of everything else in our Solar System, galaxy, and universe. Everything is interactive, interdependent, and interrelated. The solar radiation that reaches Earth is dissipated in one of the following ways: by reflection, by absorption, and by scattering. Solar radiation may return to outer space, be absorbed/scattered in the atmosphere, or be absorbed by the Earth's surface. As radiation is scattered in the atmosphere. A new concept of correlation of sunspots, Sweeping Earth directed Coronal Mass Ejections from Sun are being done by the Environmental Scientists globally. No one can doubt that the Sun is the chief driving force for our terrestrial climate. The annual March of the seasons as the Earth's axis of rotation tilts toward or away from the Sun's direction is sufficient proof of that, while the presence of periodicities in glacial deposits matching those of known orbital variations has revealed the apparent sensitivity of global climate to relatively small changes in the distribution of sunlight. What has remained debatable and controversial, however, is the question of whether or not variations in the Sun's radiative and plasma emissions occur that are capable of influencing the weather and climate at the Earth's surface. Judith Lean and David Rind of NASA has first attempted to correlate the Sun with the climate, whose

S. Mukherjee, *Extraterrestrial Influence on Climate Change*, SpringerBriefs in Environmental Science, DOI: 10.1007/978-81-322-0730-6_5,

interests are in modeling the sensitivity of the atmosphere and climate to different forcing. Solanki, Schussler, and Fligge of Max Plank Institute, Germany has published their findings in Nature Journal in 2000. It states that modeled reconstruction of the solar magnetic field "provides the major parameter needed to reconstruct the secular variation of the cosmic ray flux impinging on the terrestrial atmosphere," as a stronger solar magnetic field "more efficiently shields the Earth from cosmic rays." The significance of this accomplishment, they say, is indicated by the fact that it has been suggested, "cosmic rays affect the total cloud cover of the Earth and thus drive the terrestrial climate." Their work should thus enable this proposed Sun-climate link to be more thoroughly and rigorously examined. A brief discussion is given in this article on influence of star flares on Sun-Earth environment. Decrease in magnetic values and electron flux are also noticed after the Earth directed star storm (NASA 2004). These geophysical factors seem to have some relationships with the snowfall in near pole latitudes and higher altitudes and movement of clouds and development of fog, smog, and rainfall on 25 December 2004.

But although the Sun is known to be a variable star, its total output of radiation is often assumed to be so stable that we can neglect any possible impacts on climate. Testimony to this assumption is the term that has been employed for more than a century to describe the radiation in all wavelengths received from the Sun: the so-called "solar constant," whose value at the mean Sun-Earth distance is a little over 1 1/3 kW per square meter of surface.

In truth, the solar "constant" varies. Historical attempts to detect possible changes from the ground were thwarted by variable absorption in the air overhead. Sunspots and other forms of solar activity are produced by magnetic fields, whose changes also affect the radiation that the Sun emits, including its distribution among shorter and longer wavelengths. The Earth has a magnetic field with north and south poles. The magnetic field of the Earth is enclosed in a region surrounding the Earth called the magnetosphere. As the Earth rotates, its hot core generates strong electric currents that produce the magnetic field. This field reaches 36,000 miles into space. The magnetosphere prevents most of the particles from the Sun, carried in solar wind from impacting the Earth. The star storm distorts the shape of the magnetosphere by compressing it at the front and causing a long tail to form on the side away from the Sun. This long tail is called the magneto tail star storm and Sun storm can enter through magnetic shield and disturb atmosphere.

In the latter part of the nineteenth century, there were many claims of newfound connections between sunspots and climate. It began with the announcement by the amateur astronomer Heinrich Schwabe in 1843 that sunspots come and go in an apparently regular 11-year cycle. What followed was a flood of reported correlations, not only with local and regional weather but with crop yields, human health, and economic trends. These purported connections—that frequently broke down under closer statistical scrutiny—lacked the buttress of physical explanation and were in time forgotten or abandoned.

After more than a century of controversy, the debate as to whether solar variability has any significant effect on the climate of the Earth remains to be settled. Present work attempts to establish a new hypothesis on star-Sun-Earth atmospheric interactions and opens a new horizon of near accurate weather prediction research. Bjorck and colleagues propose that a weakening of solar activity may have caused this mini chill. It coincided; they find, with a large increase in the amount of beryllium-10 trapped in Greenland ice—evidence of a solar flicker. This radioactive form of beryllium is produced when cosmic rays from space collide with nitrogen and oxygen atoms in the atmosphere. The magnetic field around the Earth protects the planet from cosmic rays. This field is stronger when the Sun is more active—emitting more ultraviolet radiation and displaying more sunspots—so fewer cosmic rays can penetrate. Star-Sun interaction on Earth's atmosphere: For hailstorm or snowstorm or heavy cloud formation it is essential that Earth's atmosphere should have sudden increase in micro dust particles. Recent hailstorm has developed subsequently after the star storm. Since early 1992, Ulysses has been monitoring the stream of stardust flowing through our Solar System. The stardust is embedded in the local galactic cloud through which the Sun is moving at a speed of 26 km every second. As a result of this relative motion, a single dust grain takes 20 years to traverse the Solar System. Observations by the DUST experiment on board Ulysses have shown that the stream of stardust is highly affected by the Sun's magnetic field. In the 1990s, this field, which is drawn out deep into space by the out-flowing solar wind, kept most of the stardust out. The most recent data, collected up to the end of 2002, shows that this magnetic shield has lost its protective power during the recent solar maximum. In an upcoming publication in the Journal of Geophysical Research ESA scientist Markus Landgraf and his co-workers from the Max-Planck-Institute in Heidelberg report that about three times more stardust is now able to enter the Solar System. The reason for the weakening of the Sun's magnetic shield is the increased solar activity, which leads to a highly disordered field configuration. In the mid-1990s, during the last solar minimum, the Sun's magnetic field resembled a dipole field with well-defined magnetic poles (North positive, South negative), very much like the Earth. Unlike Earth, however, the Sun reverses its magnetic polarity every 11 years. The reversal always occurs during solar maximum. How the Earth's surface temperature adjusts to a given change in star-solar radiation depends on the processes by which the climate system responds to variations in the energy it receives. Some of these so-called feedbacks amplify the effects of changes that are imposed; others reduce them. Lumped together, they make up what is called the sensitivity of the climate system: the number of degrees that the mean-surface temperature will be raised or lowered in response to a given change, up or down, in solar and/or extraterrestrial radiation, or any other climate driver. To understand the impacts of star-solar variations on climate we need to know how much the star-solar inputs vary, and how the climate system responds to these changes. The sensitivity of climate to solar radiation changes, as defined earlier, is not well known. A conservative estimate is that a 0.1 % change in solar total radiation will bring about a temperature response of 0.06–0.2 °C, providing the change persists long enough for the climate system to adjust. This could take 10–100 years. Changes in visible and infrared solar radiation alter the surface temperature by simple heating; other parts of the spectrum can also affect climate, through paths that are less direct. We know, for example, that the enhanced UV radiation that pours outward from the Sun at times of high solar activity increases the amount of ozone in the stratosphere. At times of minima in the 11-year cycle, less ozone is found. One consequence of these solar perturbations is to complicate the detection of human-induced depletion of the protective ozone layer; another may be to perturb the temperature at the Earth's surface, through connections that link the upper and lower parts of the atmosphere. Solar radiation received at the Earth can vary by means that are unrelated to any changes on the Sun itself. The best studied of these are very long-term changes in the Earth's orbit around the Sun, which alter the distribution of sunlight both geographically and seasonally. They are now believed to trigger the coming and going of the major Ice Ages. As such, they may provide a powerful demonstration of the impacts of changes in solar radiation on the climate system. The changes involved arise from gradual shifts in the shape and orientation of the Earth's orbit around the Sun, and in the present 23 1/2° tilt of the Earth's axis of rotation. These cyclic changes, brought about by the changing gravitational pull of the other planets and the Moon, introduce periods of about 19, 23, 41, and 100 thousand years in the distribution of sunlight over the globe. The total annual dosage, averaged over the entire surface, varies by up to 0.1 %, while more specific, seasonal changes at any place can reach a few percent. Such changes are apparently sufficient to trigger major changes in climate—implying that the Earth's climate system may be more sensitive to small solar radiative perturbations than one might think. Climate simulations are as yet unable to account for the unexpectedly prominent 100 thousand year periodicity in the record of past climate. This long period is associated with the eccentricity of the Earth's orbit, which oscillates between circular and slightly elliptical. Accompanying changes in the Sun-Earth distance directly affect the amount of solar radiation incident on the Earth in different parts of the year. Changes in activity on the Sun itself could exert a similar effect. Such studies of solar perturbations can serve the broader cause as diagnostic probes of the atmosphere and climate system. Ambiguities regarding projected greenhouse warming call in much the same way for clearer information regarding the role of the Sun, as a possibly important contributor to the current warming trend. Climate simulations using only greenhouse gas changes predict a warming that exceeds the 0.5 °C that is documented in the instrumental record of the past 140 years. To reconcile the difference between the observed and the predicted values, either the models are wrong or other, natural or anthropogenic forcing must be properly factored in. If variations in the output of the Sun are indeed limited to the tenth of a percent that is recorded in direct measurements, future solar changes will likely have but a small effect, one way or the other, on the surface warming of a few degrees that is expected to result from doubled concentrations of greenhouse gases. If we include reasoned deductions from what we know of the Sun and climate in the past, we must allow that solar changes could potentially alter the anticipated effects of carbon dioxide and other greenhouse gases on the surface temperature of the Earth. In January 2002, a moderately dim star in the constellation Monoceros, the Unicorn, suddenly became 600,000 times more luminous than our Sun. This made it temporarily the brightest star in our Milky Way. The light from this eruption created a unique phenomenon known as a 'light echo' when it reflected off dust shells around the star. This phenomenon was followed by hailstorm in northern hemisphere. Further, in the month of December 2004 the star in the constellation repeated similar phenomenon. It may be concluded that the sudden snowfall in northern continents on 25 December 2004 has sufficient bearing on star-Sun-Earth's atmosphere interaction. It has been attempted to estimate the impact of solar variability on climate through the study of the total solar irradiance and galactic cosmic rays. The Sun is one of the main source of energy for maintaining our climate. We are yet to learn much about the way aerosols, and other gases affect regional and global climate. During the Solar eclipse the energy source reduces temporarily. Concentration of atmospheric gases, cloud and aerosol concentration depends on the variation of the energy from the Sun (Masmoudi et al. 2003). During the total solar eclipse on 23 July 2009 it was possible to correlate some atmospheric parameters like cloud cover, SO_2, NO_2, and aerosol concentration variation. The longest solar eclipse of the century cast a wide shadow for several minutes over Asia and Pacific Ocean. Satellite data of a part of China has been selected for the study of variation of atmospheric parameters during the solar eclipse on 23 July 2009.

Star-Sun influence on the Earth's atmosphere is now established. Sometimes it is long-term change otherwise it has been observed as episodic events.

For hailstorm, snowstorm, or heavy cloud formation, it is essential that the Earth's atmosphere should contain enough micro aerosol particles to act as cloud condensation nuclei. Data of the star storm show that hailstorms have developed subsequently after the star storm (NASA 2004). Since early 1992, Ulysses has been monitoring the stream of stardust flowing through our Solar System. The stardust is embedded in the local galactic cloud through which the Sun is moving at a relative speed of 26 km per second. Because of this relative motion, a single dust grain takes 20 years to traverse the Solar System. Observations by the DUST experiment onboard Ulysses has shown that the stream of stardust is highly affected by the Sun's magnetic field. In most of the 1990s, this field, which was drawn out deep into space by the out-flowing solar wind, kept most of the stardust out. The most recent data, collected up to the end of 2002, show that this magnetic shield has lost its protective power during the recent solar maximum. It has been reported that about three times more stardust is now able to enter the Solar System (Max Plank News release 1999). The reason for the weakening of the Sun's magnetic shield is the increasing solar activity, which leads to a highly disordered field configuration. In the mid-1990s, during the last solar minimum, the Suns magnetic field resembled a dipole field with well-defined magnetic poles (North positive, South negative), very much like the Earth. Unlike Earth, however, the Sun reverses its magnetic polarity every 11 years. The reversal always occurs during solar maximum. That is when the magnetic field is highly disordered, allowing more interstellar dust to enter the Solar System. It is of interest to note that in the reversed configuration after the recent solar maximum (North negative, South positive), the interstellar dust is even channeled more efficiently toward the inner Solar System. It is expected that more interstellar dust will occur from 2005 onwards, but it had appeared in December 2004. The Sun has entered the zodiac's 13th house: An interstellar wind hit our planet. It is a helium-rich breeze from the stars, flowing into the Solar System from the direction of Ophiuchus (NASA 2004). The Sun's gravity focuses the material into a cone and Earth passes through it during the first weeks of December. Earth was inside the cone during 25 December 2004. While grains of stardust are very small, about one-hundredth the diameter of a human hair, they do not directly influence the planets of the Solar System. However, the dust particles move very fast, and produce large numbers of fragments when they impact asteroids or comets. It is, therefore, conceivable that an increase for dust in the Solar System will create more cosmic dust by collisions with asteroids and comets. We know from measurements by high-flying aircraft that around 40,000 tones of interplanetary dust enters the Earth's atmosphere each year. It is possible that the increase of stardust in the Solar System will influence the amount of extraterrestrial material that rains down to Earth (ESA Science News 2003).

How the Earth's surface temperature adjusts to a given change in solar radiation depends on the processes by which the climate system responds to variations in the energy it receives. Some of these factors amplify the effects of changes that are imposed; others reduce them. Lumped together, they make up what is called the sensitivity of the climate system, which indicates the number of degrees by which the mean-surface temperature will be raised or lowered in response to a given change, up or down, in solar and/or extraterrestrial radiation or any other climate driver. To understand the impacts of star-solar variations on climate we need to know how much the star-solar inputs vary, and how the climate system responds to these changes. The sensitivity of climate to solar radiation changes, as defined earlier, is not well known. A conservative estimate is that a 0.1 % change in solar total radiation will bring about a temperature response of 0.06–0.2 °C, providing that change persists long enough for the climate system to adjust. This could take 10–100 years. Changes in the visible and infrared part of the solar spectrum alter the surface temperature by simple heating; other parts of the spectrum can also affect climate, although their paths of influence are less direct. We know, for example, that the enhanced UV radiation that pours outward from the Sun at times of high solar activity increases the amount of ozone in the stratosphere through increased dissociation of molecular oxygen. At times of minima in the 11-year cycle of the Sun, ozone is decreased. It has been also known that ozone contributes to climatic change. Solar radiation received at the Earth can vary by means that are unrelated to any changes on the Sun itself. The best studied of these are very long-term changes in the Earth's orbit around the Sun, which alter the distribution of sunlight both geographically and seasonally. They are now believed to trigger the coming and going of the major Ice Ages. As such, they may provide a powerful demonstration of the impacts of changes in solar radiation on the climate system.

The changes involved arise from gradual shifts in the shape and orientation of the Earth's orbit around the Sun, and in the present 231/2° tilt of the Earth's axis of rotation. These cyclic changes, brought about by the changing gravitational pull of the other planets and the Moon, introduce periods of about 19, 23, 41, and 100 thousand years in the distribution of sunlight over the globe. The total annual dosage, averaged over the entire surface, varies by up to 0.1 %, while more specific, seasonal changes at any place can reach a few percent. Such changes are apparently sufficient to trigger major changes in climate, hence implying that the Earth's climate system may be more sensitive to small solar irradiance perturbations than one might think. Climate simulations are as yet unable to account for the unexpectedly prominent 100 k year periodicity in the record of past climate. This long period is associated with oscillations in the eccentricity of the Earth's orbit. Changes in the Sun-Earth distance directly affect the amount of solar radiation incident on the Earth in different parts of the year. Changes in the activity of the Sun itself could exert a similar effect. Such studies of solar perturbations can serve the broader cause as diagnostic probes of the atmosphere and climate system. Ambiguities regarding projected greenhouse warming call in much the same way for clearer information regarding the role of the Sun as a possibly important contributor to the current warming trend. Climate simulations using only greenhouse gas changes predict a warming that exceeds 0.5 °C as documented in the instrumental record of the past 140 years. The reason behind difference between the observed and the predicted values may be that not all natural and anthropogenic forcing are considered in the models. If variations in the output of the Sun are indeed limited to the tenth of a percent that is recorded in direct measurements, future solar changes will likely have but a small effect, one way or the other, on the surface warming of a few degrees that is expected to result from doubled concentrations of greenhouse gases. If we consider Sun-Earth-climate connections observed in the past, we may think that star flares could potentially alter the anticipated effects of carbon dioxide and other greenhouse gases on the surface temperature of the Earth.

In January 2002, a moderately dim star in the Monoceros constellation, the Unicorn suddenly became 600,000 times more luminous than our Sun. This made it temporarily the brightest star in the Milky Way. The light from this eruption created a unique phenomenon known as a 'light echo' when it reflected off dust shells around the star. This phenomenon was followed by hailstorm in northern hemisphere. Further, in the month of December 2004, the star in the constellation repeated a similar phenomenon. It may be concluded that the sudden snowfall in northern hemisphere continents on 25 December 2004 has sufficient bearing on star-Sun-Earth's atmosphere interaction.

Sun-Earth environment Kp (planetary indices), proton flux, and electron flux exhibit changes. Sudden changes in these parameters may influence the environment of the Earth abruptly. If an E-flux rise is responsible for global warming, then an E-flux lowering may lead to snowfall. On 22 December 2004, a sudden fall in the electron flux was recorded by the SOHO satellite. Widespread snowfall was recorded in United Kingdom on 25 December 2005. A subsequent rise of the

Table 5.1 Snowfall in UK
on 25 December 2004

Location	Snowfall temperature (°C)
Aberdeen	Yes 1 degree minimum
London	No
Birmingham	Yes
Manchester	Yes
Cardiff	Yes
Belfast	Yes
Crossby	Yes 0 degree minimum
Woodford	Yes 0 degree minimum

Source Meteorological Office United Kingdom and BBC weather News

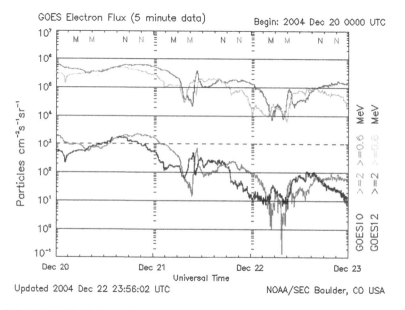

Fig. 5.1 Sudden fall of electron flux 36 h before snowfall SOHO satellite data (*Source* NASA-ESA SOHO EIT. www.spaceweather.com)

E-flux normalized the condition. The star flare might have influenced the E-flux and thus cased snowfall on 25 December 2004. Similar observations were noticed in other parts of the world also. Widespread snowfall was recorded in other parts of the world on the 25 of December 2005 and further on the 23 of February 2005 (Table 5.1). The temperature data of west USA shows a declining trend of −0.68 °F/decade, which is considered to be anomalous in the rising trend of 0.05 °F/decade for USA. Similar phenomenon has been observed in other parts of the Earth (USA and Japan) in the month of January and February 2006 followed by a star flare. Regular monitoring of star flares and their influence on the Sun-Earth environment may lead to more accurate weather prediction (Figs. 5.1 and 5.2) (Table 5.2).

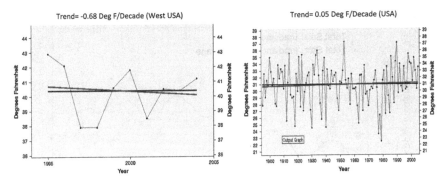

Fig. 5.2 December 1995–2005 Temperature of West region USA shows a decline while whole USA temperature since 1895 shows rising trend. Actual average and trend of temperature is shown by *red, black,* and *green* (*Source* http://climvis.ncdc.noaa.gov/)

Table 5.2 Sudden snowfall on 25 December 2004 and 23 February 2005

Location	Snowfall/cyclone	Anomalous	Remarks
1. Liverpool (UK)	Snowfall	Yes	Low Kp and low E-flux
2. Birmingham (UK)	Snowfall	Yes	-Do-
3. Manchester (UK)	Snowfall	Yes	-Do-
4. Cardiff (UK)	Snowfall	Yes	-Do-
5. Aberdeen (UK)	Snowfall	Yes	-Do-
Belfast	Snowfall	Yes	-Do-
Crossby	Snowfall	Yes	-Do-
Woodford	Snowfall	Yes	-Do-
Houghton, Michigan USA	Very high snowfall	Yes	-Do-
Hawaii, Mauna Loa, USA	Very high snowfall	Yes	-Do-
Boston, USA	Very high snowfall	Yes	-Do-
New York, USA	Very high snowfall	Yes	-Do-
Queensland, Australia	Cyclone with cold wave	Yes	-Do-
Jammu Kashmir, Shimla, India	Snowfall with cold wave	Yes	-Do-
Tehran, Iran	Snowfall	Yes	-Do-

Source Meteorological Office (UK)

What climate change is happening to other planets in the Solar System?

This argument is part of a greater one that other planets are warming. If this is happening throughout the Solar System, clearly it must be the Sun causing the rise in temperatures—including here on Earth.

It is curious that the theory depends so much on sparse information—what we know about the climates on other planets and their history—yet its proponents resolutely ignore the most compelling evidence against the notion. Over the last 50 years, the Sun's output has decreased slightly: it is radiating less heat. We can measure the various activities of the Sun pretty accurately from here on Earth, or from orbit above it, so it is hard to ignore the discrepancy between the facts and the skeptical argument that the Sun is causing the rise in temperatures (Fig. 5.3).

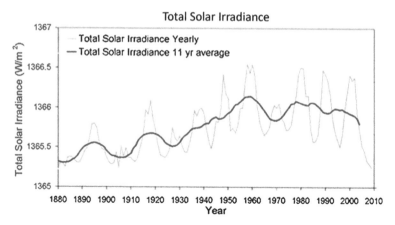

Fig. 5.3 TSI from 1880 to 1978 from Solanki (2000). TSI from 1979 to 2009 from PMOD

It is concluded that the climate change is not the only function of the carbon dioxide emission but the extraterrestrial influence has a major role to play in this. If the Sun's output has reduced, leveled off or even diminished, then what is causing other planets to warm up? Are they warming at all? The question is being asked by the supporters of global cooling. The probable answer is hidden within the changes in the cosmic ray variability from extragalactic origin and its influence on solar dynamo to fluctuate its energy balance.

Eclipse and other celestial phenomena will occur in future also, it will be interesting to monitor the changes in the environment of the Earth during these astronomical changes. There will be 36 solar eclipses from 2001 to 2025 of which 15 will be total eclipses on some part of Earth's surface—a little less than average of 1 year. A detailed study of environment is required during these eclipses to understand the climate change.

Bibliography

Arctic Climatology Project (2000) Environmental working group arctic meteorology and climate atlas. Fetterer F, Radionov V (eds) National Snow and Ice Data Center, Boulder

Bjorck S et al (2001) High-resolution analyses of an early Holocene climate event may imply decreased solar forcing as an important climate trigger. Geology 29:1107–1110

Bochnicek J, Hejda MP (2005) The winter NAO pattern changes in association with solar and geomagnetic activity. J Atmos Solar Terr Phys 67(1–2):17–32. http://jove.geol.niu.edu/faculty/stoddard/JAVA/luminaries.html

Boersma KF, Eskes HJ, Veefkind JP, Brinksma EJ, Van der ARJ, Sneep M, Van den Oord GHJ, Levelt PF, Stammes P, Gleason JF, Bucsela EJ (2007) Near real time retrieval of tropospheric NO_2 from OMI. Atm Chem Phys 2013–2128, sref:1680-7324/acp/2007-7-2103

Brandeis G, Marsh (1989) The convective liquids in a solidifying magma chamber; a fluid dynamic investigation. Nature 339:613–616

Chapman S, Ferraro VCA (1930) A new theory of magnetic storms. Nature 126:129

Cosmology (2002) Nove and supernove http://rst.gsfc.nasa.gov/Sect20/A6.html

Donn WL (1965) Meteorology. McGraw-Hill Book Company Inc, New York

ESA Science News (2003) http://sci.esa.int. Accessed 01 Aug 2003

Forbush decreases of galactic cosmic rays (1995) J Atmos Solar Terr Phys 57:1349–1355

Lean J, Beer J, Bradley RS (1995) Reconstruction of solar irradiance since 1610: implications for climate change. Geophys Res Lett 22:3195–3198

Kastner JH, Richmond M, Grosso N, Weintraub DA, Simon TA, Frank A, Hamaguchi K, Ozawa H, Henden A (2004) An X-ray outburst from the rapidly accreting young star that illuminates McNeil's nebula. Nature 430:429–431

Kokubun S, McPherron RL, Russell CT (1977) Triggering of sub storms by solar wind discontinuities. J Geophys Res 82:74

Lassen K, Friis E (1995) Christensen. J Atmos Terr Res 57:835–844

Lean J, Rind D (1996) The sun and climate. Consequences 2(1):27–36

Lean J, Beer J, Bradley R (1995) Reconstruction of solar irradiance since 1610: Implications for climate change. Geophys Res Lett 22:3195–3198

Lean J (2000) Evolution of the sun's spectral irradiance since the maunder minimum. Geophys Res Lett 27(16):2425–2428

Marris E (2006) Glacial pace picks up Greenland's ice is breaking up at an increasing rate. Published online: Nature 16 Feb 2006. doi:10.1038/news060213-11

Marsh ND, Svensmark H (2000) Low cloud properties influenced by cosmic rays. Phys Rev Lett 85(23):5004–5007

S. Mukherjee, *Extraterrestrial Influence on Climate Change*,
SpringerBriefs in Environmental Science, DOI: 10.1007/978-81-322-0730-6,
© The Author(s) 2013

Masmoudi M, Chaabane M, Medhioub K, Elleuch F (2003) Variability of aerosol optical thickness and atmospheric turbidity in Tunisia. Atmos Res 66(3):175–188

Max Plank New release, Ulysses measures the deflection of galactic dust particles by solar radiation (1999) (http://www.mpg.de/)

Mukherjee S, Mukherjee A (2002) Seeking links between Solar activity and Monsoon. J Geo You 2(7):18–23 website http://www.geographyandyou.com

Mukherjee S (2001) Space based early warning system to understand Seismotectonics. Geo Surv India Spec Publ. No.65 (II) 2001:39–44

Mukherjee S (2008) Cosmic influence on sun-earth environment. Sensors 8:7736–7752. doi: 10.3390/s8127736 www.mdpi.com/journal/sensors

Mukherjee S (2003) 26th January 2001 earthquake of Gujarat, India was triggered by change in Kp and electron flux induced by Sun. In: Proceedings of international workshop on earth system processes related to Gujarat earthquake using space technology. January 27–29, IIT Kanpur, India, http://home.iitk.ac.in/~ramesh/3days_schedule.doc

Mukherjee S, Mukherjee A (2002) UV-B flux increase during coronal mass ejection. In: TIGER 4th (Virtual) thermospheric/ionospheric geospheric research (TIGER) symposium on Long-term measurement of solar EUV/UV fluxes for thermospheric/ionospheric modelling and for space weather investigations website http://worktools.si.umich.edu Workshop on Internet

Mukherjee S (2002) Solar maximum has changed environment of the earth. In: Proceedings living with a star science workshop (13–15 Nov 2002), NASA, USA. Website: http://lws.gsfc.nasa.gov

Mukherjee S, Mukherjee A (2002) Change in magnetic field: an early warning system to understand Seismotectonics. In: Strassemeir KG, Washuettl A (eds) Proceedings of 1st Potsdam thinksop on sunspots and starspots, pp.139–142. AIP, Potsdam, Germany. Website: www.aip.de

NASA news, a breeze from the star. NASA news of 17 Dec 2004

NEIC (2001) http://neic.usgs/gov/neis/bulletin/010213142208.html

Olson EL, Allen RM (2005) Nature 438:212–215

Pudovkin MI, Veretenenko SV (1995) Cloudiness decreases associated with Forbush-decreases of galactic cosmic rays. J Atmos Terr Phys 57:1349

Rignot E, Kanagaratnam P (2006) Changes in the velocity structure of the greenland Ice sheet. Science 311:986–990. doi:10.1126/science.1121381

Roychoudhury P (1999) Int J Mod Phys A14 1961

Solanki SK, Schussler M, Fligge M (2000) Evolution of the sun's large-scale magnetic field since the Maunder minimum. Nature 408:445–447

Srivastava HN (1983) Forecasting earthquakes. National Book Trust, New Delhi, p 11

Tinsley BA, Dean GW (1991) Apparent tropospheric response to MeV–GeV particle flux variations: a connection via electro-freezing of supercooled water in high-level clouds? J Geophys Res 96:22283–22296

Yu F (2004) Formation of large NAT particles and denitrification in polar stratosphere: possible role of cosmic rays and effect of solar activity. Atmos Chem Phys Discuss 4:1037–1062

Zesta E, Singer HJ, Lummerzheim D, Russell CT, Lyons LR, Brittnacher MJ (2000) The effect of the January 10, 1997, pressure pulse on the magnetosphere–ionosphere current systems. Geophys Monogr Ser, vol 118. Ohtani S-I, Fujii R, Hesse M, Lysak RL (eds) AGU, Washington